THE THEORY OF
TOPOLOGICAL
SEMIGROUPS

MONOGRAPHS AND TEXTBOOKS IN
PURE AND APPLIED MATHEMATICS

1. *K. Yano*, Integral Formulas in Riemannian Geometry (1970)*(out of print)*
2. *S. Kobayashi*, Hyperbolic Manifolds and Holomorphic Mappings (1970) *(out of print)*
3. *V. S. Vladimirov*, Equations of Mathematical Physics (A. Jeffrey, editor; A. Littlewood, translator) (1970) *(out of print)*
4. *B. N. Pshenichnyi*, Necessary Conditions for an Extremum (L. Neustadt, translation editor; K. Makowski, translator) (1971)
5. *L. Narici, E. Beckenstein, and G. Bachman*, Functional Analysis and Valuation Theory (1971)
6. *D. S. Passman*, Infinite Group Rings (1971)
7. *L. Dornhoff*, Group Representation Theory (in two parts). Part A: Ordinary Representation Theory. Part B: Modular Representation Theory (1971, 1972)
8. *W. Boothby and G. L. Weiss (eds.)*, Symmetric Spaces: Short Courses Presented at Washington University (1972)
9. *Y. Matsushima*, Differentiable Manifolds (E. T. Kobayashi, translator) (1972)
10. *L. E. Ward, Jr.*, Topology: An Outline for a First Course (1972) *(out of print)*
11. *A. Babakhanian*, Cohomological Methods in Group Theory (1972)
12. *R. Gilmer*, Multiplicative Ideal Theory (1972)
13. *J. Yeh*, Stochastic Processes and the Wiener Integral (1973) *(out of print)*
14. *J. Barros-Neto*, Introduction to the Theory of Distributions (1973) *(out of print)*
15. *R. Larsen*, Functional Analysis: An Introduction (1973) *(out of print)*
16. *K. Yano and S. Ishihara*, Tangent and Cotangent Bundles: Differential Geometry (1973) *(out of print)*
17. *C. Procesi*, Rings with Polynomial Identities (1973)
18. *R. Hermann*, Geometry, Physics, and Systems (1973)
19. *N. R. Wallach*, Harmonic Analysis on Homogeneous Spaces (1973) *(out of print)*
20. *J. Dieudonné*, Introduction to the Theory of Formal Groups (1973)
21. *I. Vaisman*, Cohomology and Differential Forms (1973)
22. *B.-Y. Chen*, Geometry of Submanifolds (1973)
23. *M. Marcus*, Finite Dimensional Multilinear Algebra (in two parts) (1973, 1975)
24. *R. Larsen*, Banach Algebras: An Introduction (1973)
25. *R. O. Kujala and A. L. Vitter (eds.)*, Value Distribution Theory: Part A; Part B: Deficit and Bezout Estimates by Wilhelm Stoll (1973)
26. *K. B. Stolarsky*, Algebraic Numbers and Diophantine Approximation (1974)
27. *A. R. Magid*, The Separable Galois Theory of Commutative Rings (1974)
28. *B. R. McDonald*, Finite Rings with Identity (1974)
29. *J. Satake*, Linear Algebra (S. Koh, T. A. Akiba, and S. Ihara, translators) (1975)

30. *J. S. Golan*, Localization of Noncommutative Rings (1975)
31. *G. Klambauer*, Mathematical Analysis (1975)
32. *M. K. Agoston*, Algebraic Topology: A First Course (1976)
33. *K. R. Goodearl*, Ring Theory: Nonsingular Rings and Modules (1976)
34. *L. E. Mansfield*, Linear Algebra with Geometric Applications: Selected Topics (1976)
35. *N. J. Pullman*, Matrix Theory and Its Applications (1976)
36. *B. R. McDonald*, Geometric Algebra Over Local Rings (1976)
37. *C. W. Groetsch*, Generalized Inverses of Linear Operators: Representation and Approximation (1977)
38. *J. E. Kuczkowski and J. L. Gersting*, Abstract Algebra: A First Look (1977)
39. *C. O. Christenson and W. L. Voxman*, Aspects of Topology (1977)
40. *M. Nagata*, Field Theory (1977)
41. *R. L. Long*, Algebraic Number Theory (1977)
42. *W. F. Pfeffer*, Integrals and Measures (1977)
43. *R. L. Wheeden and A. Zygmund*, Measure and Integral: An Introduction to Real Analysis (1977)
44. *J. H. Curtiss*, Introduction to Functions of a Complex Variable (1978)
45. *K. Hrbacek and T. Jech*, Introduction to Set Theory (1978)
46. *W. S. Massey*, Homology and Cohomology Theory (1978)
47. *M. Marcus*, Introduction to Modern Algebra (1978)
48. *E. C. Young*, Vector and Tensor Analysis (1978)
49. *S. B. Nadler, Jr.*, Hyperspaces of Sets (1978)
50. *S. K. Segal*, Topics in Group Rings (1978)
51. *A. C. M. van Rooij*, Non-Archimedean Functional Analysis (1978)
54. *L. Corwin and R. Szczarba*, Calculus in Vector Spaces (1979)
53. *C. Sadosky*, Interpolation of Operators and Singular Integrals: An Introduction to Harmonic Analysis (1979)
54. *J. Cronin*, Differential Equations: Introduction and Quantitative Theory (1980)
55. *C. W. Groetsch*, Elements of Applicable Functional Analysis (1980)
56. *I. Vaisman*, Foundations of Three-Dimensional Euclidean Geometry (1980)
57. *H. I. Freedman*, Deterministic Mathematical Models in Population Ecology (1980)
58. *S. B. Chae*, Lebesgue Integration (1980)
59. *C. S. Rees, S. M. Shah, and C. V. Stanojević,* Theory and Applications of Fourier Analysis (1981)
60. *L. Nachbin*, Introduction to Functional Analysis: Banach Spaces and Differential Calculus (R. M. Aron, translator) (1981)
61. *G. Orzech and M. Orzech*, Plane Algebraic Curves: An Introduction Via Valuations (1981)
62. *R. Johnsonbaugh and W. E. Pfaffenberger*, Foundations of Mathematical Analysis (1981)
63. *W. L. Voxman and R. H. Goetschel*, Advanced Calculus: An Introduction to Modern Analysis (1981)
64. *L. J. Corwin and R. H. Szcarba*, Multivariable Calculus (1982)
65. *V. I. Istrătescu*, Introduction to Linear Operator Theory (1981)
66. *R. D. Järvinen*, Finite and Infinite Dimensional Linear Spaces: A Comparative Study in Algebraic and Analytic Settings (1981)

67. *J. K. Beem and P. E. Ehrlich*, Global Lorentzian Geometry (1981)
68. *D. L. Armacost*, The Structure of Locally Compact Abelian Groups (1981)
69. *J. W. Brewer and M. K. Smith, eds.*, Emmy Noether: A Tribute to Her Life and Work (1981)
70. *K. H. Kim*, Boolean Matrix Theory and Applications (1982)
71. *T. W. Wieting*, The Mathematical Theory of Chromatic Plane Ornaments (1982)
72. *D. B. Gauld*, Differential Topology: An Introduction (1982)
73. *R. L. Faber*, Foundations of Euclideañ and Non-Euclidean Geometry (1983)
74. *M. Carmeli*, Statistical Theory and Random Matrices (1983)
75. *J. H. Carruth, J. A. Hildebrant, and R. J. Koch*, The Theory of Topological Semigroups (1983)
76. *R. L. Faber*, Differential Geometry and Relativity Theory: An Introduction (1983)
77. *S. Barnett*, Polynomials and Linear Control Systems (1983)
78. *G. Karpilovsky*, Commutative Group Algebras (1983)
79. *F. Van Oystaeyen and A. Verschoren*, Relative Invariants of Rings: The Commutative Theory (1983)
80. *I. Vaisman*, A First Course in Differential Geometry (1984)
81. *G. W. Swan*, Applications of Optimal Control Theory in Biomedicine (1984)
82. *T. Petrie and J. D. Randall*, Transformation Groups on Manifolds (1984)
83. *K. Goebel and S. Reich*, Uniform Convexity, Hyperbolic Geometry, and Nonexpansive Mappings (1984)
84. *T. Albu and C. Năstăsescu*, Relative Finiteness in Module Theory (1984)
85. *K. Hrbacek and T. Jech*, Introduction to Set Theory, Second Edition, Revised and Expanded (1984)
86. *F. Van Oystaeyen and A. Verschoren*, Relative Invariants of Rings: The Noncommutative Theory (1984)
87. *B. R. McDonald*, Linear Algebra Over Commutative Rings (1984)
88. *M. Namba*, Geometry of Projective Algebraic Curves (1984)
89. *G. F. Webb*, Theory of Nonlinear Age-Dependent Population Dynamics (1985)
90. *M. R. Bremner, R. V. Moody, and J. Patera*, Tables of Dominant Weight Multiplicities for Representations of Simple Lie Algebras (1985)
91. *A. E. Fekete*, Real Linear Algebra (1985)
92. *S. B. Chae*, Holomorphy and Calculus in Normed Spaces (1985)
93. *A. J. Jerri*, Introduction to Integral Equations with Applications (1985)
94. *G. Karpilovsky*, Projective Representations of Finite Groups (1985)
95. *L. Narici and E. Beckenstein*, Topological Vector Spaces (1985)
96. *J. Weeks*, The Shape of Space: How to Visualize Surfaces and Three-Dimensional Manifolds (1985)
97. *P. R. Gribik and K. O. Kortanek*, Extremal Methods of Operations Research (1985)
98. *J.-A. Chao and W. A. Woyczynski, eds.*, Probability Theory and Harmonic Analysis (1986)
99. *G. D. Crown, M. H. Fenrick, and R. J. Valenza*, Abstract Algebra (1986)
100. *J. H. Carruth, J. A. Hildebrant, and R. J. Koch*, The Theory of Topological Semigroups, Volume 2 (1986)

Other Volumes in Preparation

THE THEORY OF
TOPOLOGICAL
SEMIGROUPS

Volume 2

J. H. CARRUTH

University of Tennessee
Knoxville, Tennessee

J. A. HILDEBRANT
R. J. KOCH

Louisiana State University
Baton Rouge, Louisiana

MARCEL DEKKER, INC.　　　　　New York and Basel

Library of Congress Cataloging-in-Publication Data
(Revised for volume 2)

Carruth, J. Harvey (James Harvey), [date]
    The theory of topological semigroups.

    (Monographs and textbooks in pure and applied
mathematics ; 75, 100.)
    Bibliography: v. 1, p.
    Includes index.
    1. Topological semigroups. I. Hildebrant, J. A.
(John A.), [date]. II. Koch, R. J., [date]. III. Title.
IV. Series: Monographs and textbooks in pure and
applied mathematics ; 75, 100.
QA387.C36 1983   512'.55   82-23560
ISBN 0-8247-1795-3 (v. 1)
ISBN 0-8247-7320-9

MARCEL DEKKER, INC.
270 Madison Avenue, New York, New York 10016

Current printing (last digit):
10 9 8 7 6 5 4 3 2 1

PRINTED IN THE UNITED STATES OF AMERICA

Preface

Cohomology has been one of the more useful tools in the study of topological semigroups since the work of A. D. Wallace in the early 1950s. The first chapter of this book discusses cohomology results and their applications to compact monoid theory. Additional applications are contained in the section on admissibility in Chapter 5.

One of the most researched areas of topological semigroups has been that of topological semilattices. Chapter 2 gives an extensive discussion of this topic, including theorems on Lawson semilattices and continuous semilattices.

Chapter 3 discusses the results on irreducible semigroups of Hofmann and Mostert, Hunter, and Koch. It employs the result that the group of units of a compact divisible semigroup is connected, and it states (without proof) and uses the Hofmann and Mostert transformation group fixed point theorem.

Results in Chapter 4 deal with the translational hull of a topological semigroup and its applications to the theory of ideal extensions. Included in this chapter is a simplified proof that the translational hull of a compact reductive semigroup is again compact.

The discussion of semigroups on continua in Chapter 5 is mainly a mixture of older topics in topological semigroups (from the 1950s and 1960s), such as admissibility, with some results on these same topics from the 1970s. Some useful constructions are detailed and their applications are discussed. The chapter is concluded with results of a number of contributors on the two-cell.

Chapter 6 on semigroup categories will be of interest to those who have some familiarity with category theory and wish to study topological semigroups from this point of view. Categories considered in this chapter include topological semigroups, topological groups, and topological semilattices, and the subcategories of the compact objects of these categories.

Chapter 7, on manifold semigroups, contains earlier material on affine semigroups from the 1960s and more modern material on Lie semigroups from the 1980s.

The final chapter consists entirely of a list of over 50 unsolved problems from a variety of sources and some remarks on existing partial solutions.

In order to make this book more self-contained, three appendixes are provided. The topics are cohomology, topological spaces, and category theory.

The authors wish to express their appreciation to Carol Collins for her many useful remarks regarding the preparation of the manuscript for this book.

J. H. Carruth
J. A. Hildebrant
R. J. Koch

Contents

PREFACE   iii

CONTENTS OF VOLUME I   ix

1   COHOMOLOGY AND COMPACT MONOIDS   **1**

Basic Cohomology Structure Theorems   1
Peripherality   5

2   TOPOLOGICAL SEMILATTICES   **10**

Basic Concepts   10
Lawson Semilattices   12
Separating Homomorphisms   16
A Non-Lawson Semilattice   18
Arc Chains   23
Continuous Semilattices   24
Breadth of a Semilattice   28

3   IRREDUCIBLE SEMIGROUPS   **35**

Power Set Semigroups   35
Irreducible Semigroups Are Abelian   37
An Irreducible Semigroup Has a Trivial Group of Units   41
·Koch's Thread Theorem   45

**4 THE TRANSLATIONAL HULL AND IDEAL
    EXTENSIONS**   **49**

Basic Construction   49
Extensions of Bitranslations   54
Divisibility   56
Computational Techniques   58
Finite Semigroups   61
Ideal Extensions   66

5 SEMIGROUPS ON CONTINUA   **72**

Friedberg's Constructions   72
Admissibility   74
Semigroups on Ruled Continua   84
Semigroups on the Two-Cell   89

6 SEMIGROUP CATEGORIES   **98**

7 MANIFOLD SEMIGROUPS   **115**

Affine Semigroups   115
Lie Semigroups   123

8 PROBLEMS   **135**

Wallace's Problems   135
The Problems of Hofmann and Mostert   138
The Compact Semilattice Problems of Brown
    and Stralka   141
Problems of Koch   144
Problems on Divisible Semigroups   145
Translational Hull Problems   147
Problems of Carruth and Clark   147

APPENDIX A   COHOMOLOGY   **149**

APPENDIX B   TOPOLOGICAL SPACES   **161**

Contents

APPENDIX C   CATEGORY THEORY   **169**

BIBLIOGRAPHY   **181**

SYMBOLS   **192**

INDEX   **193**

Contents of Volume 1

1 FUNDAMENTAL CONCEPTS

Semigroups
Subsemigroups
Subgroups
Ideals
Homomorphisms and Congruences

2 NEW SEMIGROUPS FROM OLD

Cartesian Products
Semidirect Products
Rees Products
Adjunction Semigroups
Projective and Injective Limits
Semigroups of Homomorphisms
Semigroup Compactifications
Free Topological Semigroups
Coproducts

3 INTERNAL STRUCTURE THEOREMS

Monothetic Semigroups
Completely Simple Semigroups
Green's Relation

The Multiplicative Structure of a $\mathcal{D}$-Class
Schützenberger Groups
Green's Quasi-Orders

4 I-SEMIGROUPS

Topological Preliminaries
Basis I-Semigroups
The Structure of I-Semigroups
Closed Congruences on I-Semigroups

5 ONE PARAMETER SEMIGROUPS

Divisibility and Connectedness
Existence of One-Parameter Semigroups
Solenoidal Semigroups

6 COMPACT DIVISIBLE SEMIGROUPS

The Set of Divisible Elements
The Exponential Function
Cone Semigroups
Subunithetic Semigroups
Finite Dimensional Semigroups
Further Results on Divisible Semigroups

SYMBOLS

BIBLIOGRAPHY

THE THEORY OF
TOPOLOGICAL
SEMIGROUPS

1
Cohomology and Compact Monoids

In this chapter we consider applications of cohomology theory to the theory of compact monoids. This includes a study of the concepts of middle and edge points, and conditions one can impose on a compact monoid to make it a group.

BASIC COHOMOLOGY STRUCTURE THEOREMS

In this section we establish results pertaining to the cohomological structure of a compact monoid. It will be proved that a compact monoid has the same cohomology as its minimal ideal.

1.1 Theorem Let S be a topological semigroup which contains a locally compact minimal ideal M. Then the inclusion map $i:M \to S$ induces a surmorphism $i^*:H^*(S) \to H^*(M)$.

Proof: In view of 3.23(c) of Carruth, Hildebrant, and Koch [1983], we see that M is a retract of S. The result follows from Appendix A.7. ∎

1.2 Theorem Let S be a compact connected semigroup with a left identity 1. Then $H^*(S)$ is isomorphic to $H^*(M(S))$.

Proof: Now S acts on (S, M) by multiplication $m: S \times (S, M)$ $\rightarrow (S, M)$, where $M = M(S)$. Thus, by the generalized homotopy lemma (see Appendix A.27), $\hat{s}^* = \hat{t}^*: H^*(S, M) \rightarrow H^*(S, M)$ for each $s, t \in S$. Fix $k \in M$. Then the diagram

$$(S,M) \xrightarrow{\bar{k}} (M,M) \xrightarrow{i} (S,M)$$
$$\xrightarrow[\hat{k}]{}$$

where $\bar{k}(x) = kx$, i is inclusion, and $\hat{k}$ is defined by composition, induces the diagram

$$H^*(S,M) \rightarrow H^*(M,M) \rightarrow H^*(S,M)$$
$$\xrightarrow[\hat{k}^*]{}$$

Since $H^*(M,M) = 0$, $\hat{k}^*$ is the 0 map. Now 1^* is the identity, and hence $H^*(S,M) = 0$. From the exact sequence of Appendix A.4:

$$\swarrow$$
$$H^n(S,M) \rightarrow H^n(S) \rightarrow H^n(M)$$
$$\swarrow$$
$$H^{n+1}(S,M) \rightarrow$$

we obtain that $H^n(S)$ is isomorphic to $H^n(M)$ for each $n \in \mathbf{N}$. In case $n = 0$, $H^n(S) = 0 = H^n(M)$, since both S and M are connected. ∎

An alternate proof of 1.2 employing the Rees quotient may be of interest: S/M acts on S/M by multiplication, and the identity homomorphism is $\hat{1}^* = \hat{0}^* = 0$, so that $H^n(S/M) = 0$. An exact sequence argument yields that $H^n(S)$ is isomorphic to $H^n(M)$.

We turn now to a discussion of the position of the minimal ideal in a compact monoid. As we have seen, if S is a topological semigroup with a locally compact minimal ideal, then $M(S)$ is a kind of generalized deformation of S. In addition we also have:

1.3   Theorem   If S is a compact connected monoid, then S is locally connected at $M(S)$; i.e., each open set containing $M(S)$ contains an open connected set which contains $M(S)$.

Proof: Let U be an open set containing $M(S)$. Then, by 1.31(b) and 1.27 of Carruth, Hildebrant, and Koch [1983], $J_0(U)$ is open and connected, and $M(S) \subseteq J_0(U) \subseteq U$. ∎

Under certain conditions the minimal ideal of a compact monoid S can be determined by the cohomology of S.

1.4 Theorem Let S be a compact connected monoid such that $cd\, S = n$ and $H^n(S) \neq 0$. Then $M(S)$ is a group and is a unique floor for each $h \in H^n(S)$ such that $h \neq 0$.

Proof: Let $h \in H^n(S)$ with $h \neq 0$, and let F be a floor for h. Suppose that for some $t \in S$, $F \cap (S \setminus tF) \neq \square$. Then $tF \cap F$ is a proper subset of F, so that $cd(SF) > n$, which contradicts $cd\, S = n$. We obtain that $F \subseteq tF$ for each $t \in S$, and by the swelling lemma, $tF = F$ for each $t \in S$. It follows that F is a minimal left ideal of S and similarly, F is a minimal right ideal such that $F = Ft$ for each $t \in S$. Finally, we see that $F = M(S)$ is a group. ∎

The minimal ideal in the setting of 1.4 has the topological property of being invariant under homeomorphisms of S into itself.

1.5 Corollary Let S be a compact connected monoid such that $cd\, S = n$ and $H^n(S) \neq 0$, and let $\phi\!:\!S \to S$ be a homeomorphism of S onto S. Then $\phi(M(S)) = M(S)$.

The result of 1.2 has proved useful in investigating which spaces can be the underlying space of a compact connected monoid. We present a pair of examples to illustrate this. Recall that $\mathbf{N}$ denotes the discrete additive semigroup of positive integers, $\mathbf{Z}$ denotes the discrete group of integers, $\mathbf{R}$ denotes the additive group of real numbers with the usual topology, $\mathbf{H}$ denotes the additive semigroup of nonnegative real numbers with the usual topology, and $I_u = [0,1]$ with the usual topology and multiplication.

Example   Let e denote the identity of the circle group **R**/**Z** and let S be the compact subsemigroup of **R**/**Z** $\times I_u$ consisting of (**R**/**Z** $\times$ $\{0\}$) $\cup$ ($\{e\} \times I_u$). Then S is a compact monoid on a circle with an interval $I = [a,b]$ attached at a:

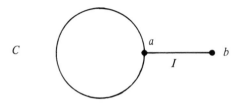

Observe that $C = M(S)$ is a group and that b is the identity of S.

Example   Let $R = \{(a,b) \in \mathbf{H} \times \mathbf{H} : 5/2 \le a,\, 5/2 \le b,\, \text{and}\, |a-b| \in$ **N**$\} \cup \triangle$ (**H**). Then R is a closed congruence on **H** and $S = \mathbf{H}/R$ is a compact monoid on a circle C with an interval $I = [a,\, b]$ attached at a. Again, as in the previous example, we have that $C = M(S)$ is a group and b is the identity of S.

Suppose that S is a compact connected monoid whose underlying space is a circle C with an interval I attached at a. We will show that the positions of $M(S)$ and the identity 1 are precisely as in the examples above [namely, $M(S) = C$ and $1 = b$], and that $M(S)$ is a group.

Now if $1 \in M(S)$, then $M(S) = S$ is a group. Since S is not homogeneous, we conclude that $1 \notin M(S)$, so that, by virtue of 1.2, $C \subseteq M(S)$, $1 \in I$ and $1 \ne a$.

We will show that $1 = b$. Suppose, for the purpose of proof by contradiction, that $1 \ne b$. Then 1 is a cutpoint of S. Let $S \setminus \{1\} = A \cup B$, where A is the component of $S \setminus \{1\}$ containing a and B is the component of $S \setminus \{1\}$ containing b. Then for $t \in B$, we have $t \in t\bar{A}$, since $1 \in \bar{A}$. Moreover, $t\bar{A} \cap M(S) \ne \square$, since $M(S) \subseteq \bar{A}$. It follows that $1 \in t\bar{A}$, and $t \in H(1)$. Thus, $B \subseteq H(1)$. But $H(1)$ is homogeneous and B contains cutpoints of $H(1)$ and the noncutpoint b. This contradiction yields $1 = b$.

The conclusions that $C = M(S)$ is a group now follows from 1.2.

We note in this illustration that the identity 1 is in a peripheral position on S. This concept will be investigated in the next section.

1.6 Theorem Let S be a compact connected monoid and let $e \in M(S) \cap E(S)$. Then $H^*(S)$ is isomorphic to $H^*(H(e))$.

Proof: Define $r:S \to H(e)$ by $r(x) = exe$ and let $i:H(e) \to S$ be inclusion. Define $F: S \times S \to S$ by $F(x, t) = txt$. Then $F(x, e) = exe = ri(x)$ for each $x \in H(e)$ and $F(x, 1) = 1x1 = x = 1_S(x)$ for each $x \in S$. It follows that $H(e)$ is a deformation retract of $S[H(e) = eSe$, since $e \in M(S) \cap E(S)]$, and hence $H^*(S) \approx H^*(H(e))$. ∎

PERIPHERALITY

The cohomological notion of peripherality has proved to be a useful tool in compact monoid theory. One type of application is in determining whether certain continua admit the structure of a compact monoid, and another is in establishing that some conditions imposed on a compact monoid force it to be a group. We will discuss the latter application in this section, and defer the former application to a later chapter.

Two concepts of cohomological peripherality of interest appear in the literature. In Hofmann and Mostert [1966] points of a regular space which we refer to as edge points were called "marginal points" and middle points were called "intrinsic points." Related material appears in Wallace [1953] and Bredon [1967]. In Lawson and Madison [1970a] points of a space which we refer to as 0-edge points were called "peripheral points" and 0-middle points were called "inner points." In that paper the authors present a comparison of the two concepts of peripherality and treat compact monoids in a sequel [Lawson and Madison 1970b].

We will restrict our attention to the edge point concept and applications, and will mention briefly the 0-edge point notion.

If X is a space and $p \in X$, the p is called an *edge point* of X provided that for each open set U containing p, there exists an open set V such that $p \in V \subseteq U$ and inclusion $i: X \backslash V \hookrightarrow X$ induces an isomorphism $i^*:H^*(X) \to H^*(X \backslash V)$. We say that p is a *middle* *point* of X otherwise. The set of edge points of X is called the *edge* of X and is denoted *edge*(X), and the set of middle points of X is called the *middle* of X and is denoted *middle*(X).

The notion of middle point is of a local character. If $p \in X$, then $p \in middle(\underline{X})$ if there exists an open set W containing p such that $p \in middle(\overline{W})$. This follows from the excision theorem:

$$H^*(X, X \backslash V) \approx H^*(X \backslash (X \backslash \overline{W}), \ (X \backslash V) \backslash (X \backslash \overline{W}))$$

$$= H^*(\overline{W}, \ \overline{W} \backslash V)$$

Observe also that *edge*(X) is defined independently of the way in which X may be embedded in any other space. The terms applied to these notions are hereditary and intrinsic. Indeed, they are local. If p and q have homeomorphic neighborhoods with the homeomorphism taking p to q, then p is an edge point if and only if q is an edge point.

It was observed in Wallace [1953] that if X is a compact subset of $\mathbf{R}^n$, then *edge*$(X) \subseteq \partial X$, where $\partial X = X \cap \overline{\mathbf{R}^n \backslash X}$ (the boundary of X in $\mathbf{R}^n$).

The first result which we present concerning edge points of a monoid appears in Hofmann and Mostert [1966]. It employs the fact that an irreducible subsemigroup of a compact connected monoid meets the group of units only in the identity. We will establish this result in a later chapter.

1.7   Theorem   If S is a compact connected monoid which is not a group, then $1 \in edge(S)$.

Proof: Let U be an open set such that $1 \in U \subseteq S \backslash M(S)$, and let T be a compact connected subsemigroup of S such that $T \cap M(S) \neq \square$ and $T \cap H(1) = \{1\}$ (an irreducible subsemigroup of S has this property). Let $V = S \backslash T(S \backslash U)$ and note that $V \subseteq U$ and $T(S \backslash V) = S \backslash V$. Now fix $k \in T \cap M(S)$ and let $\hat{k}:(S, S \backslash V) \to (S, S \backslash V)$ be defined by $\hat{k}(x) = kx$. Thus T acts on $(S, S \backslash V)$, so that $\hat{k}^* = 1_S^*$ by the generalized homotopy theorem. But $\hat{k}(S) \subseteq$

$M(S) \subseteq S \setminus V$, and hence $\hat{k}^* = 0$. Finally, we see that $H^*(S, S \setminus V)$ $= 0$ and $H^*(S) \approx H^*(S \setminus V)$ (see Appendix A.4). ∎

1.8 Corollary   If S is a compact connected monoid which is not a group, then $H(1) \subseteq edge(S)$.

Proof: Let $h \in H(1)$. Then $\lambda_h : S \to S$ is a homomorphism of S onto S such that $\lambda_h(1) = h$. Since $1 \in edge(S)$, we have $h \in edge(S)$. ∎

The existence of middle points is implicit in Bredon [1967] with the argument here appearing in Lawson and Madison [1970].

1.9   Theorem   A finite dimensional continuum has a middle point.

Proof: Let X be a continuum with $cd\, X = n$. Then $cd\, X \not\le n - 1$, so there exist a closed set $A \subseteq X$ with $H^{n-1}(X) \to H^n(A)$ (induced by inclusion) which is not surjective. Thus there exists $h \ne 0$, $h \in H^{n-1}(A)$, such that h is not extendible to X. Let R be a roof for h. Then h is not extendible to $R \cup A$, but h is extendible to $T \cup A$ if T is a proper closed subset of R. Let $x \in R \setminus A$. We will show that x is a middle point of X. Suppose the contrary. Let U be an open set containing x. Then there exists an open set V with $x \in V \subseteq U$ and $V \cap A = \square$. Let $W = S \setminus V$, $B = W \cup R \cup A$ and consider the diagram

$$H^{n-1}(W) \oplus H^{n-1}(R \cup A) \to H^{n-1}((R \cup A) \setminus V) \to H^n(B) \to H^n(W) \oplus H^n(R \cup A)$$

$$\uparrow \alpha \qquad\qquad\qquad \uparrow \beta \qquad\qquad \uparrow \lambda \qquad\qquad \uparrow \delta$$

$$H^n(S) \oplus H^{n-1}(R \cup A) \to H^{n-1}(R \cup A) \to H^n(S) \to H^n(S) \oplus H^n(R \cup A)$$

where the vertical maps are induced by inclusion and the horizontal maps are Meyer–Vietoris diagrams. Since x is assumed to be an edge point, the maps α and δ are isomorphisms. Since

$cd\ X = n$, γ is surjective, and hence β is surjective. In view of this and the diagram:

$$H^{n-1}(R \cup A) \xrightarrow{\beta} H^{n-1}((R \cup A)\backslash V) \to H^{n-1}(A)$$

we see that h is extendible to $H^{n-1}(R \cup A)$, contradicting that R is a roof for h. ∎

The following result due to Hudson and Mostert [1963] is an important application of these theorems.

1.10 Theorem If S is a compact connected finite dimensional homogeneous monoid, then S is a group.

Proof: If S is not a group, then by 1.7, $1 \in edge(S)$, and since S is homogeneous, each point of S is an edge point. This contradicts 1.9. ∎

We also have:

1.11 Theorem If S is a compact connected monoid such that 1 has a euclidean neighborhood, then S is a group.

Related results appear in Madison [1969], Madison and Selden [1967], and Selden [1961]. A particularly noteworthy result is that if S is a coset space of a compact connected group and if S is a monoid, then S is a group.

A point p in a space X is called a 0-*edge point* of X provided that for each open set U containing p, there exists an open set V with $p \in V \subseteq U$ such that the inclusion $j:(X,X\backslash V) \to (X,X\backslash U)$ induces the 0-map $j^*:H^*(X,X\backslash U) \to H^*(X,X\backslash V)$. The set of 0-edge points of X is called the 0-*edge* of X and is denoted $0\text{-}edge(X)$. If $q \in X$ is not a 0-edge point of X, the q is called a 0-*middle point* of X, and the set of 0-middle points of X is called the 0-*middle of* X and denoted $0\text{-}middle(X)$.

Observe that for a space X, we have that $p \in edge(X)$ if and only if $H^*(X, X \setminus V) = 0$ for cofinally many neighborhoods of p. It follows that $edge(X) \subseteq 0-edge(X)$, and hence $0- \neq middle(X) \subseteq middle(X)$.

For a compact connected monoid S which is not a group, an argument similar to that of 1.7 can be employed to show that $1 \in 0-edge(S)$.

In Lawson and Madison [1970b] a generalization of 1.10 is obtained by replacing "homogeneous" with "locally homogeneous" in the hypothesis.

2
Topological Semilattices

In this chapter we present an introduction to the theory of topological semilattices. Most of this material can be found in Lawson [1969] and Lawson [1970a]. Some results from Gierz, Hofmann, Keimel, Lawson, Mislove, and Scott [1980] and Brown and Stepp [1982] are discussed. We discuss Lawson semilattices both in terms of their properties and in terms of separating homomorphisms into the min interval. Some of the results stem from earlier results in topological semilattices [Brown, 1965], lattices [Birkhoff, 1967], and topological lattices [Strauss, 1968]. Further results in topological semilattice theory appear in Anderson and Ward [1961], Borrego [1970], Bowman [1974], Brown and Stralka [1973, 1977], Crawley [1976], Friedberg [1972], Hofmann and Stralka [1973a, 1973b], Lau [1972a, 1972b, 1973a, 1973b, 1975] Lawson, Liukkonan, and Mislove [1977], Lawson and Williams [1970], Lea [1972, 1976], Lea and Lau [1975], Rhodes [1973], Stepp [1971, 1973, 1975a, 1975b], Stralka [1977], Wallace [1961], and Williams [1975]. Related material on semilattices and pospaces appears in Koch [1959, 1960, 1965].

BASIC CONCEPTS

Recall that a *topological semilattice* is an abelian topological semigroup S such that each element of S is idempotent. Observe that a subsemigroup of a semilattice is a semilattice (and is therefore called a subsemilattice), a cartesian product of semilattices is a semilattice, and a projective limit of compact semilattices is a compact semilattice.

If we consider the min interval I_m, then any nonempty subset of I_m is a subsemilattice, e.g., the Cantor set. An alternate version of the semilattice on the Cantor set can be obtained by considering the two point min semilattice $\{0,1\}$ and taking a countable cartesian product of this semilattice with itself.

If S is a topological semilattice and we define $e \leq f$ for $e, f \in S$ provided $ef = e$, then $\leq$ is a closed partial order on S called the *natural order*. It is, of course, the $\mathcal{H}$ order on S. For $e, f \in S$, we have $inf\{e, f\} = ef$, so that the function $(e,f) \mapsto inf\{e, f\}$ is a continuous function from $S \times S$ to S. If, on the other hand, one begins with a Hausdorff space X endowed with a partial order such that $inf\{x, y\}$ exists for each $x, y \in X$ and $(x,y) \mapsto inf\{x, y\}$ is continuous, then X with the multiplication defined by $xy = inf\{x, y\}$ is a topological semilattice.

If S is a semilattice and $x \in S$, we write $\downarrow x = \{a : a \leq x\}$, $\uparrow x = \{a : x \leq a\}$, and for $A \subseteq S$, we denote $\downarrow A = \cup \{ \downarrow x : x \in A\}$ and $\uparrow A = \cup \{ \uparrow x : x \in A\}$ (see Appendix B).

If S is a topological semilattice ($\leq$ is the natural order on S), then for $e \in S$, we have $\downarrow e = Se$, and e is an identity for $\downarrow e$.

2.1   Theorem   Let S be a compact semilattice and let A be a nonempty subset of S. Then
  (a) S has a zero element 0.
  (b) If A is bounded above, then *sup A* exists.
  (c) *inf A* exists.
  (d) If A is closed, then $\downarrow A$ and $\uparrow A$ are closed.

Proof: To prove (a) observe that S is down directed and hence S has an *inf* 0 (see Appendix B.4). For the proof of (b), let U be the set of upper bounds of A. Let $e, f \in U$. Then $p \leq e$ and $p \leq f$ for each $p \in A$, and hence $p^2 = p \leq ef$, and $ef \leq f$, so that U is down directed. We conclude that $inf\ U = a$ exists (see Appendix B.4), and we have $a = sup\ A$. To prove (c), observe that the set L of lower bounds of A is nonempty, since $0 \in L$. Since L is bounded above by A, $b = sup\ L$ exists by (b), and $b = inf\ A$. Now (d) follows from the fact that $\leq$ is a closed subset of $S \times S$. ∎

We prove a special case of the Wallace acyclicity theorem [Wallace, 1961a]. Recall that a space X is acyclic provided $H^*(X) = 0$.

2.2 Theorem Let S be a compact connected semilattice. Then S is acyclic.

Proof: For $e \in S$, $\downarrow e = eS$ is a compact connected monoid with 0, and hence $H^*(eS) = 0$ from 1.2. For $A \subseteq S$, $\downarrow A = \cup \{eS : e \in A\}$, and since eS is connected and $0 \in eS$, we have that $\downarrow A$ is connected. From Appendix A.26, we have $H^*(S) = H^*(\downarrow S) = 0$. ∎

LAWSON SEMILATTICES

For a topological semilattice S and $e \in S$, we say that S has *small semilattices* at e provided e has a neighborhood base of subsemilattices of S. Observe that if the space S is regular and S has small semilattices at e, then e has a base of closed semilattice neighborhoods, since the closure of a subsemilattice is a subsemilattice.

If S is a semilattice which has small semilattices at each of its points, then S is called a *Lawson semilattice*. In this section we discuss sufficient conditions for a semilattice to be a Lawson semilattice.

2.3 Lemma Let S be a locally compact semilattice and let $x \in S$. If $\downarrow x$ has small semilattices at x, then S has small semilattices at x.

Proof: Assume that $\downarrow x$ has small semilattices at x, and let U be an open set containing x such that $\overline{U}$ is a compact partially ordered space and hence is locally convex from Appendix B.6.

Since $x^2 = x$, there exists an open set V containing x such that $V^2 \subseteq U$ and V is convex in $\bar{U}$.

Define $\phi: S \to \downarrow x$ by $\phi(s) = sx$. Then ϕ is a continuous surmorphism of S onto $\downarrow x$. Let K be a subsemilattice of $\downarrow x$ which is a neighborhood of x in $\downarrow x$ and which is contained in $V \cap \downarrow x$. Then $\phi^{-1}(K)$ is a subsemilattice of S and a neighborhood of x in S.

To complete the proof, we set $A = \phi^{-1}(K) \cap V$ and show that A is a subsemilattice of S. Let $a, b \in A$. Then $ab \in \phi^{-1}(K) \cap V^2$, and $abx \in \phi^{-1}(K) \cap \downarrow x = K \subseteq V$. Since $abx \leq ab \leq a$, abx and a are elements of V, and V is convex in $\bar{U}$, we conclude that $ab \in V$. It follows that $ab \in A$. ∎

It is immediate from 2.3 that any locally compact semilattice with a zero 0 has small semilattices at 0. This result is useful in reducing the problem of showing that a locally compact semilattice S is a Lawson semilattice to the problem of assuming that S has an identity 1 and showing that S has small semilattices at 1.

It is straightforward to show that a subsemilattice of a Lawson semilattice is a Lawson semilattice, and that the cartesian product of Lawson semilattices is a Lawson semilattice. Along this line, we have the following:

2.4   Theorem   Let S be a compact Lawson semilattice and let $\phi: S \to T$ be a continuous surmorphism of S onto a compact semilattice T. Then T is a Lawson semilattice.

Proof: We first assume that T has an identity 1, and show that T has small semilattices at 1. Let W be an open neighborhood of 1 and let V be an open convex subset of T such that $1 \in V \subseteq W$. Note that $\phi^{-1}(1)$ is a closed subsemilattice of S, and let z be the least element of $\phi^{-1}(1)$. Since S is a Lawson semilattice, there exists a neighborhood K of z such that $K \subseteq \phi^{-1}(V)$ and K is a subsemilattice of S. Then $\downarrow K^\circ$ is an open set containing $\phi^{-1}(1)$. Since ϕ is closed, $\phi(\downarrow K^\circ)$ is a neighborhood of 1.

If $t \in \uparrow(K)$, then $r \leq t$ for some $r \in K$, and $\phi(r) \leq \phi(t) \leq 1$. Since V is convex, $\phi(t) \in V$, and hence $\phi(\uparrow K) \subseteq V$. It follows that $\phi(\uparrow K)$ is a subsemilattice of T containing 1 and contained in W.

We now reduce the general case to the case just completed. Let $y \in T$. Since $\downarrow y$ is closed, $\phi^{-1}(\downarrow y)$ is a compact semilattice. Now $\phi | \phi^{-1}(\downarrow y)$ is a continuous homomorphism and y is an identity for $\downarrow y$. The conclusion now follows from 2.3. ∎

Observe that if S is a topological semilattice and the natural order $\leq$ is a total order on S, then S is a Lawson semilattice, since each nonempty subset of S is a subsemilattice. Employing this fact and 2.4, we can generate a large class of examples of Lawson semilattices.

The following result will be useful. It is due to Wallace and appears in Koch [1957].

2.5 Lemma   Let S be a locally compact monoid such that the component C of 1 is compact and let V be an open set containing C. Then there exists an open and compact subsemigroup T of S such that $C \subseteq T \subseteq V$.

Proof: Let W be an open and compact set such that $C \subseteq W \subseteq V$, and let $T = \{x \in S : xW \subseteq W\}$. Then T is a subsemigroup of S containing 1, $T \subseteq W$, and since W is both open and compact, so is T. Since C is connected, we have $C \subseteq T$. ∎

2.6 Theorem   Let S be a locally compact totally disconnected semilattice. Then S is a Lawson semilattice.

Proof: If $x \in S$, then $\downarrow x$ is a locally compact totally disconnected semilattice with identity x. From 2.5, we see that $\downarrow x$ has small semilattices at x. The conclusion follows from 2.3. ∎

2.7 Lemma   Let S be a topological semigroup and I a closed ideal of S such that $\overline{S \backslash I}$ is compact. Then S/I is a compact semigroup.

Proof: Let $U = S \backslash I$, and let $B = \overline{U}$. Then S/I is homeomorphic to $B/\partial(B)$, where $\partial(B)$ is the boundary of B, and hence S/I is compact. Let $\pi : S \to S/I$ be the natural homomorphism. If $(\pi(x), \pi(y)) \in S/I \times S/I$, then continuity of multiplication at $(\pi(x), \pi(y))$ is clear if neither x nor y is in 1. Consider a pair of the form $(\pi(x), \pi(I))$. Then $\pi(x)\,\pi(I) = \pi(I)$. Let W be an open set with $I \subseteq W$ and let $V = \{x \in S: B^I \times B^I \subseteq W\}$. Then V is an open ideal with $I \subseteq C \subseteq W$. Let P be an open set with $x \in P$. Then $PV \subseteq W$. ∎

2.8 Lemma Let S be a compact connected finite dimensional semilattice with identity 1 such that S is locally connected at 1. Then there exists $x \in S$ such that $x \neq 0$ and $1 \in (\uparrow x)^\circ$ (interior of $\uparrow x$).

Proof: Let $n = cd\,S$. Then there exists a closed subset B of S and $h \in H^{n-1}(B)$ such that h is not extendible to S. We can assume that there exists $y \in B$ such that $h|\{y\} = 0$. This is true for any $y \in B$ if $n-1 \geq 1$. If $n-1 = 0$, we let $y \in S\backslash B$, redefine h to be $(h,0) \in H^0(B) \times H^0(\{y\}) \equiv H^0(B \cup \{y\})$, and redefine B to be $B \cup \{y\}$. To continue, let R be a roof for h and let $x \in R\backslash(B \cup \{0\})$. Such an x exists, since h is extendible to $B \cup \{0\}$. Since $x \notin B$, there exists a closed connected neighborhood N of 1 such that $x \notin NB$.
  We claim that $N \subseteq \uparrow x$. Let $z \in N$, and let $A = B \cup \{0\}$ and $Z = N \cup \downarrow z$. Since S is acyclic by 2.2, Z is a continuum, and $A \subseteq ZA$. From Appendix A.25 we obtain that $R \subseteq ZA$. It follows that $x \in Z(B \cup \{0\}) = ZB \cup \{0\}$. Since $x \neq 0$, we have $x \in ZB = (N \cup \downarrow z)B = NB \cup \downarrow zB$. From $x \notin NB$, we have $x \in (\downarrow z)B$, i.e., $x = ab$, for $a \in \downarrow z$, $b \in B$. Hence, $x \leq a \leq z$ and $z \in \uparrow x$. ∎

2.9 Theorem. Let S be a locally compact locally connected finite dimensional semilattice. Then S is a Lawson semilattice.

Proof: Let $u \in S$. Then the component C of u also satisfies the hypothesis of the Theorem. Since C is open, S has small semilattices at u if and only if C has small semilattices at u. Hence, we may assume, without loss of generality, that S is connected.
  Now $\downarrow u$ is a locally connected compact finite dimensional connected semilattice, since $\downarrow u$ is a retract of S. Let V and W be an open connected neighborhood of u in $\downarrow u$ such that $\bar{V}$ is compact

and $W^2 \subseteq V$. Let $I = \downarrow (\bar{V}\backslash W)$. Since $\bar{V}\backslash W$ is compact, I is a closed ideal. In view of the fact that $W \subseteq V$ and $\bar{V}\backslash W \subseteq I$, we see that $W \cup I = \bar{W} \cup I$ is closed. Since $W^2 \subseteq V$, we have $W \cup I$ is a semilattice. By 2.7, $(W \cup I)/I$ is a compact semilattice. Since it is homeomorphic to $\bar{W}/\partial(\bar{W})$, it is finite dimensional from Appendix A.28 and is connected, since W is connected. Let $\phi: W \cup I \to (W \cup I)/I$ be the natural map. Then, by 2.8, there exists $x \in (W \cup I)/I$, $x \neq \phi(I)$, such that $\phi(u) \in (\uparrow x)^\circ$, so that $\phi^{-1}(\uparrow x)$ is a semilattice neighborhood of u in $\downarrow u$ contained in V. The theorem follows from 2.3. ■

SEPARATING HOMOMORPHISMS

In this section we consider conditions for which $Hom(S, I_m)$ separates points for a topological semilattice S, where $I_m = [0, 1]$ is said to be *decreasing* [*increasing*] if $\downarrow B = B[\uparrow B = B]$.

A topological semilattice S is called a *U-semilattice* if for each $x \in S$ and each open increasing subset V such that if $x \in V$ there exists $y \in V$ such that $x \in (\uparrow y)^\circ$.

2.10 Lemma. Let S be a U-semilattice, A a closed decreasing subset of S, and let $b \in S\backslash A$. Then there exists $\phi \in Hom(S, I_m)$ such that $\phi(A) = 0$ and $\phi(b) = 1$.

Proof: For each dyadic rational number ξ in $[0, 1]$ we will define an open set $V(\xi)$. Let $V(0) = \square$ and $V(1) = S\backslash \uparrow b$. Since A is decreasing, $S\backslash A$ is increasing, and hence $b \in (\uparrow r)^\circ$ for some $r \in S\backslash A$. Since $r \in S\backslash A$ and $S\backslash A$ is increasing, there exists $s \in S\backslash A$ such that $r \in (\uparrow s)^\circ$. Set $V(1/4) = S\backslash \uparrow s$. Since $(\uparrow r)^\circ$ is open, we have that $\uparrow ((\uparrow r)^\circ)$ is open. Note that $(\uparrow r)^\circ \subseteq \uparrow ((\uparrow r)^\circ) \subseteq \uparrow r$, and hence $(\uparrow r)^\circ = \uparrow ((\uparrow r)^\circ)$, i.e., $(\uparrow r)^\circ$ is increasing. Thus there exists $t \in (\uparrow r)^\circ$ such that $b \in (\uparrow t)^\circ$. Let $V(3/4) = S\backslash \uparrow t$. Continuing recursively in this manner, we define $V(\xi)$ for each dyadic rational ξ in $[0, 1]$ to satisfy:

(a) $V(\xi) = S\backslash \uparrow x$ for some $x \in S$.
(b) $\xi < \eta$ implies that $\overline{V(\eta)} \subseteq V(\eta)$.

(c) $A \subseteq V(\eta)$ if $\eta \neq 0$.
(d) $V(0) = \square$ and $V(1) = S \backslash \uparrow b$.

Define $\phi : S \to [0, 1]$ by $\phi(x) = inf\{\xi : x \in V(\xi)\}$ if $x \in S \backslash \uparrow b$, and $\phi(x) = 1$ if $x \in \uparrow b$. As in the proof of Urysohn's lemma, ϕ is continuous, $\phi(A) = 0$, and $\phi(\uparrow b) = 1$. Since $b \in \uparrow b$, we have $\phi(b) = 1$.

To see that ϕ is order preserving, let $x \leq y$. Then $y \in V(\xi)$ implies that $x \in V(\xi)$ since each $V(\xi)$ is decreasing. Hence $\phi(x) \leq \phi(y)$.

It remains to show that ϕ is a homomorphism. Let $x, y \in S$ and suppose that $\phi(x) \leq \phi(y)$. Since $xy \leq x$ and ϕ is order preserving, we have $\phi(xy) \leq \phi(x)$. Now if $\phi(xy) < \phi(x)$, then $\phi(xy) < \xi < \phi(x)$ for some dyadic rational ξ. From property (b), we have $xy \in V(\xi)$ and $x \notin V(\xi)$. Now, $y \notin V(\xi)$, since $\phi(x) \leq \phi(y)$. By property (a), there exists $t \in S$ with $V(\xi) = S \backslash \uparrow t$. Since $x, y \in \uparrow t$, we have $xy \in \uparrow t$. But we already have that $xy \in S \backslash \uparrow t$. Hence the assumption that $\phi(xy) \leq \phi(x)$ leads to a contradiction. Thus $\phi(xy) = \phi(x) = \phi(x)\phi(y)$. ∎

2.11 Theorem If S is a U-semilattice, then $Hom(S, I_m)$ separates points.

Proof: Let $p, q \in S$ with $p \neq q$. We can assume that $p \not\leq q$. Thus $p \notin \downarrow q$, and $\downarrow q$ is a closed decreasing set. By 2.10, there exists $\phi \in Hom(S, I_m)$ such that $\phi(\downarrow q) = 0$ and $\phi(p) = 1$. ∎

2.12 Theorem Let S be a locally compact Lawson semilattice. Then S is a U-semilattice.

Proof: Let $x \in V$, where V is an open increasing subset of S. Then there exist neighborhoods W and K of x such that $\overline{W} \subseteq V$, $\overline{W}$ is compact, K is a closed subsemilattice, and $K \subseteq W$. Hence K is compact, and thus has a zero element z. We have that $x \in K^\circ \subseteq (\uparrow z)^\circ$. ∎

2.13 Corollary Let S be a locally compact Lawson semilattice. Then $Hom(S, I_m)$ separates points.

2.14 Corollary Let S be a compact semilattice. Then S is a Lawson semilattice if and only if S is topologically isomorphic to a subsemilattice of a product of min intervals.

A NON-LAWSON SEMILATTICE

The abundance of Lawson semilattices is apparent from 2.4, 2.6, 2.9, 2.13, and from the fact that they are both hereditary and productive. One suspects that a non-Lawson semilattice is most likely to be pathological. Indeed, this is the case. In this section we display an example of a non-Lawson semilattice. Along with the verification of its properties presented here, this example and others appear in Lawson [1970].

2.15 Lemma Let S be a compact Hausdorff space with a semilattice multiplication $m{:}S \times S \to S$ such that $\leq$ (the natural order) is closed in $S \times S$ and m is continuous at each point of $\leq$. Then m is continuous and hence S is a compact semilattice.

Proof: We write xy for $m(x,y)$ and consider nets $x_\alpha \to x$ and $y_\alpha \to y$ in S. Then $x_\alpha y_\alpha \xrightarrow{f} z$ for some $z \in S$, since S is compact. We will show that $z = xy$. Now $x_\alpha y_\alpha \leq x_\alpha$, and $\leq$ is closed, so that $z \leq x$ and hence $z \leq xy$. Now $(x_\alpha, xy) \to (y,xy) \in \leq$, and continuity of m at points of $\leq$ yields that $x_\alpha xy \to xy$ and $y_\alpha xy \to xy$. Likewise, $(x_\alpha xy, y_\alpha xy) \to (xy,xy)$ implies that $x_\alpha y_\alpha xy \to xy$. Now $(x_\alpha y_\alpha xy, x_\alpha y_\alpha) \to (xy,z)$ and $x_\alpha y_\alpha xy \leq x_\alpha y_\alpha$. Since $\leq$ is closed, we have $xy \leq z$, and finally $z = xy$. ∎

For $n \in \mathbf{N}$ with $1 < n$, let $\alpha_n = 1/m2^{m-1}$, where $2^{m-1} < n \leq 2^m$. Observe that $\sum_{n=1}^\infty \alpha_n$ is divergent.

2.16 Lemma For each $\varepsilon > 0$, there exists $p \in \mathbf{N}$ such that if $k \in \mathbf{N}$ and $p \leq k$, then

$$\sum_{n=2}^{2k} \alpha_n < \varepsilon + \sum_{n=2}^{k} \alpha_n$$

Proof: Note that if $A = \{n : 2^{m-1} < n \le 2^m\}$, then $\Sigma_{n \in A} \alpha_n = 1/m$ for each $m \in \mathbf{N}$. Choose p and q such that $2/\varepsilon < q$ and $2^{q-1} < p$. Now if $p \le k$, then there exists a unique $m \in \mathbf{N}$ such that $2^{m-1} < k \le 2^m$. Then

$$\sum_{n=2}^{2k} \alpha_n \le \sum_{n=1}^{2^{m+1}} \alpha_n = \sum_{n=2}^{2^{m-1}} \alpha_n + \frac{1}{m} + \frac{1}{m+1} \le \sum_{n=2}^{k} \alpha_n + \frac{2}{m}$$

Since $q \le m$, we have $2/m \le 2/q \le \varepsilon$. ∎

For each $i \in \mathbf{N}$, let $s(i)$ be the least integer such that $i \le \Sigma_{n=1}^{s(i)} \alpha_n$, and for each $n \in \mathbf{N}$, let $\{0, 1\}_n$ denote the two-point min semilattice. Denote $S_i = \Pi_{n=1}^{s(i)} \{0, 1\}_n$, and for $x \in S_i$ let $\theta(x)$ denote the number of zero coordinates of x. Define $\sigma_i : S_i \to [0, \infty]$ by

$$\sigma_i(x) = \begin{cases} \infty & \text{if } \theta(x) = 0 \\ i & \text{if } \theta(x) = 1 \\ s(i) & \text{if } \theta(x) = s(i) \\ i - \sum_{n=2}^{\theta(x)} \alpha_n & \text{otherwise} \end{cases}$$

where $[0, \infty]$ is the one-point compactification of the nonnegative real numbers.

2.17 Lemma   Each σ_i is an order preserving function from S_i into $[0, \infty]$. If $0 < \varepsilon < \tau$ are positive numbers, then there exists $q \in \mathbf{N}$ such that if $q \le i$, $x, y \in S_i$, $\tau < \sigma_i(x)$, and $\tau < \sigma_i(y)$, then $\tau - \varepsilon < \sigma_i(xy)$.

Proof: That each σ_i is order preserving is a straightforward consequence of its definition. Assume that $0 < \varepsilon < \tau$. For $0 < \varepsilon$, choose $p \in \mathbf{N}$ according to 2.16, and choose $q \in \mathbf{N}$ such that $\tau + \Sigma_{n=2}^{2p} \alpha_n < q$.

Suppose that $q \le i$, $x, y \in S_i$, $\tau < \sigma_i(x)$, and $\tau < \sigma_i(y)$. Let $z = xy$. We will show that $\tau - \varepsilon < \sigma_i(z)$

Now either $\theta(z) \le 2\theta(x)$ or $\theta(z) \le 2\theta(y)$, since $z = xy$ can have at most twice as many zero coordinates as one of x or y. We can assume, without loss of generality, that $\theta(z) \le 2\theta(x)$. Note that, from the definition of σ_i, in all cases $i - \Sigma_{n=2}^{\theta(z)} \alpha_n \le \sigma_i(z)$ if the summation is interpreted to be 0 for $\theta(z)$ equal 0 or 1.

If $\theta(x) \le p$, then

$$\tau \le q - \sum_{n=2}^{2p} \alpha_n \le q - \sum_{n=2}^{2\theta(x)} \alpha_n \le i - \sum_{n=2}^{\theta(z)} \alpha_n \le \sigma_i(z)$$

the first inequality follows from the choice of q. Hence if $\theta(x) \le p$, then $\tau - \varepsilon < \sigma_i(z)$.

If $p < \theta(x)$, then

$$\tau - \varepsilon < \sigma_i(x) - \epsilon = i - \left[\sum_{n=2}^{\theta(x)} \alpha_n + \varepsilon \right] \le i - \sum_{n=2}^{2\theta(x)} \alpha_n \le i$$

$$- \sum_{n=2}^{\theta(z)} \alpha_n \le \sigma_i(z)$$

Hence $\tau - \varepsilon < \sigma_i(z)$. ∎

Let $K = [0, \infty] \times \Pi_{i=1}^{\infty} S_i$, and let $T = \{(t,(x_1,x_2, \ldots)) \in K : t \le \sigma_i(x_i)$ for all $i\}$.

2.18 Lemma   The space T is compact.

Proof: Since K is compact, we need only show that T is closed in K. Let $(t,(x_1,x_2,\ldots)) \in K \backslash T$. There exists an open neighborhood U of $(t,(x_1,x_2,\ldots))$ such that if $(s,(y_1,y_1,\ldots)) \in U$, then $\sigma_n(x_n) < s$ and $y_n = x_n$; then $\sigma_n(y_n) = \sigma_n(x_n) < s$ implies $(s,(y_1,y_2,\ldots)) \in K \backslash T$. It follows that $K \backslash T$ is open, T is closed in K, and hence T is compact. ∎

We represent $(x_1,x_2,\ldots) = x$ and define multiplication on T as follows: $(s,x)(t,y) = (u,xy)$, where $u = s \wedge t \wedge inf\{\sigma_i(x_i y_i) : i \in N\}$, and $a \wedge b = min\{a, b\}$ in $[0,\infty]$.

2.19 Lemma T is a compact semilattice.

each $x \in A$, there exists a compact neighborhood
K_x is a subsemilattice. Since A is compact, some
of K_x's have interiors which cover A, i.e.,
over A. Let $K = \bigcup_{i=1}^{n} K_i$. K^n is a compact
such that $A \subseteq K \subseteq K^n$. ∎

em Let S be a locally compact Lawson semilattice
If S is arcwise connected, or if S is connected and
nected, then each element of S lies on an arc chain
0.

f: Let $x \in S$. Then either hypothesis guarantees the
of a continuum P with $0, x \in P$. Let $T = (\overline{\bigcup_{n \in \mathbf{N}} P^n})$. Then
mpact subsemilattice of S and T is connected. From 2.23, x
an arc chain containing 0. ∎

TINUOUS SEMILATTICES

rge portion of the effort in semilattice theory in the past decade
been exerted in the field of continuos semilattices. We present
re a brief view of some of the basic concepts of this theory along
th the result which is called The Fundamental Theorem of
ompact Semilattices in Gierz, Hofmann, Keimel, Lawson,
Mislove, and Scott [1980; see 2.29]. The work of Hofmann and
Stralka in 1973 appears to have marked the merging of semigroup
concepts with those of Scott to begin this era of investigation.

If S is a semilattice and $a, b \in S$ then a is said to be *way below b*
(denoted $a \ll b$) provided that for each directed subset D of S, if b
$\leq sup\ D$, then there exists $d \in D$ such that $a \leq d$. For $x \in S$, we
denote $\Downarrow x = \{a \in S : a \ll x\}$ and $\Uparrow x = \{a \in S : x \ll a\}$.

A semilattice S is said to be *up-complete* if each up-directed
set has a *sup* in S.

A semilattice S is called a *continuous semilattice* provided that
S is up-complete and for each $x \in S$, $\Downarrow x$ is directed and $x = sup\ \Downarrow x$.

The *Lawson topology* on a semilattice S is the topology on S

Proof: We first show that multiplication is associative
(commutativity is clear). Let (s, x), (t, y), and (r, z) be in T. Then
$[(s, x)(t, y)](r, z) = (u, xy)(r, z) = (v, xyz)$, where $u = s \wedge t \wedge$
$inf\{\sigma_i(x_i y_i) : i \in \mathbf{N}\}$ and $v = u \wedge r \wedge inf\{\sigma_i(x_i y_i z_i) : i \in \mathbf{N}\} = s \wedge t \wedge r \wedge$
$inf\{\sigma_i(x_i y_i) : i \in \mathbf{N}\} \wedge inf\{\sigma_i(x_i y_i z_i) : i \in \mathbf{N}\} = s \wedge t \wedge r \wedge$
$inf\{\sigma_i(y_i y_i z_i) : i \in \mathbf{N}\}$, since σ_i is order preserving. Now
$(s,x)[(t,y)(r,z)] = (s,x)(w,yz) = (q,xyz)$, where $w = t \wedge r \wedge$
$inf\{\sigma_i(y_i z_i) : i \in \mathbf{N}\}$ and $q = s \wedge w \wedge inf\{\sigma_i(x_i y_i z_i) : i \in \mathbf{N}\} = s \wedge t \wedge r$
$\wedge inf\{\sigma_i(y_i z_i) : i \in \mathbf{N}\} \wedge inf\{\sigma_i(x_i y_i z_i) : i \in \mathbf{N}\} = s \wedge t \wedge r \wedge$
$inf\{\sigma_i(x_i y_i z_i)\}$, and $[(s,x)(t,y)](r,z) = (s,x)[(t,y)(r,z)]$.

For $(s,x) \in T$, we have $(s,x)(s,x) = (u,xx) = (u,x)$, where $u = s$
$\wedge s \wedge inf\{\sigma_i(x_i) ; i \in \mathbf{N}\} = s \wedge inf\{\sigma_i(x_i) : i \in \mathbf{N}\}$. Since $(s,x) \in T, s \leq$
$\sigma_i(x_i)$ for each $i \in \mathbf{N}$ and hence $u = s$. We obtain that $(s,x)^2 = (s,x)$,
and T is a semilattice.

It remains only to show that multiplication on T is continuous.
Let (s,x) and (t,y) be in T. We will show that multiplication is
continuous at $((s,x), (t,y))$. In view of 2.15, we can assume that
$(t,y) \leq (s,x)$.

We first consider the case that $0 < t, s < \infty$. For $n \in \mathbf{N}$ and $0 <$
ε, let $3\varepsilon < t$ and let $W = \{(u,z) \in T : t - 3\varepsilon < u < t + 3\varepsilon$ and $y_i = z_i$
for $i \leq n\}$. Then W is a basic open neighborhood of y. Let $q \in \mathbf{N}$ as
guaranteed by 2.17 for $\tau = t - \varepsilon$ and ε. Now let $m = max\{n,q\}$ and
define neighborhoods U and V of x and y respectively as follows:

$$U = \{(s'.a) \in T : s - \varepsilon < s' < s + \varepsilon, a_i = x_i; i \leq m\}$$

and

$$V = \{(t'.b) \in T : t - \varepsilon < t' < t + \varepsilon, b_i = y_i; i \leq m\}$$

To complete the proof, we show that $UV \subseteq W$.

Let $(s',a) \in U$ and $(t',b) \in V$. Then $(s',a)(t',b) = (u,ab)$,
where $u = s' \wedge t' \wedge inf\{\sigma_i(a_i b_i) : i \in \mathbf{N}\}$. We obtain that $u \leq t' \leq t +$
3ε and $a_i b_i = x_i y_i = y_i$ for $i \leq n$, since $n \leq m$. Now $(t',b) \in V$, so that
$t - \varepsilon < t' \leq \sigma_i(a_i)$ for $i \in \mathbf{N}$. If $i \leq m$, then $t - 2\varepsilon < t - \varepsilon < \sigma_i(b_i) =$
$\sigma_i(a_i b_i)$. If $m < i$, then $q < i$ and $(t - \varepsilon) - \varepsilon < \sigma_i(a_i b_i)$ from 2.17.
Hence $t - 3\varepsilon < t - 2\varepsilon \leq s' \wedge t' \wedge inf\{\sigma_i(a_i b_i) : i \in \mathbf{N}\} = u$, and (u, ab)
$\in W$. ∎

Since T is homeomorphic to a closed subset of the cartesian product of the Cantor set and the unit interval, we see that T is one dimensional. Observe that $1 = (\infty, (1,1, \ldots))$ is an identity for T.

2.20   Lemma   If A is a subsemilattice of T with $1 \in A^\circ$, then $A \cap (0 \times \Pi_{i=1}^\infty S_i) \neq \square$.

Proof: At 1, T has a basis of open sets of the form: $U_j = \{(r,x) \in T : j < r, x_i = (1,1,\ldots,1) \text{ for } i \leq j\}$ for $j \in \mathbf{N}$. Assume that j is chosen so that $U_j \subseteq A^\circ$. Define $Q = \{(j+1,x) \in K : x_i = (1,1,\ldots,1)$ for $i \neq j+1$ and x_{j+1} has one zero entry$\}$. Then Q has $s(j+1)$ elements. For each $(j+1,x) \in Q$ we have $inf\{\sigma_i(x_i) : i \in \mathbf{N}\} = \sigma_{j+1}(x_{j+1}) = j+1$, and hence $Q \subseteq T$, so that $Q \subseteq U$. Let $(t,z) = inf\, Q$. Since A is a subsemilattice $(t,z) \in A$. From $(t,z) \in Q$, we have $t \leq \sigma_{j+1}(z_{j+1}) = 0$ since z_{j+1} has all zero entries. Hence $t = 0$.   ∎

Let $I = \{(0,x) \in T\}$. Then I is a closed ideal of T. Let $S = T/I$, and $\pi : T \to S$ natural.

2.21 Example S is a compact non-Lawson semilattice.

Proof: That S is a compact semilattice follows from the preceding lemmas. Assume that there exists a continuous homomorphism $\phi : S \to I_m$ which is nontrivial. Then $\phi(0) < \phi(1)$, and $\phi\pi : T \to I_m$ is a continuous homomorphism such that $\phi\pi(I) = \phi(0)$. Choose r so that $\phi(0) < r < \phi(1)$. Then $(\phi\pi)^{-1}[r,\phi(1)]$ is a neighborhood of 1 in T, a subsemilattice of T, and misses I. This contradicts 2.20, and hence no nontrivial continuous homomorphism exists.   ∎

In Anderson and Hunter [1969] the following question was posed: Is the Bohr compactification of a discrete semilattice totally disconnected? J. D. Lawson showed that this need not always be the case (unpublished). We present Lawson's argument:

Let S be a compact non-Lawson semilattice and let S_d be the semilattice S given the discrete topology. Let $\phi : S_d \to S$ be the identity function on S, and let $\beta : S_d \to B(S_d)$ denote the Bohr

compactificatio.
ism $\bar\phi : B(S_d) \to$.

commutes. If $B(S_d)$ were tot.
Lawson semilattice from 2.9,
semilattice from 2.7.

ARC CHAINS

2.22   Theorem   Let S be a compact not a group and let V be an open set contains an arc.

Proof: If there exists an open set W co. $\cap V \cap E$ has no local minimum so that V Appendix B.9). Otherwise, there exists $e \in E \cap$ that $eSe \cap E \cap V = \{e\}$. Since $e \notin M(S)$, eSe i. Hence eSe is compact connected monoid in which set containing e but not other idempotent. From 5. Hildebrant, and Koch [1983], there is a one-paramet in $eSe \cap V$ and hence it contains an arc.   ∎

2.23   Theorem   Let S be a compact connected semilattic each element of S lies on an arc chain containing 0.

Proof: For $e \in S$, $\downarrow e = Se$ and hence $\downarrow e$ is connected. result now follows from Appendix B.9.   ∎

2.24   Lemma   Let S be a locally compact Lawson semilattice and let A be a compact subset of S. Then there exists a compact subsemilattice T of S such that $A \subseteq T$.

Proof: We first show that multiplication is associative (commutativity is clear). Let (s, x), (t, y), and (r, z) be in T. Then $[(s, x)(t, y)](r, z) = (u, xy)(r, z) = (v, xyz)$, where $u = s \wedge t \wedge inf\{\sigma_i(x_iy_i):i \in \mathbf{N}\}$ and $v = u \wedge r \wedge inf\{\sigma_i(x_iy_iz_i):i \in \mathbf{N}\} = s \wedge t \wedge r \wedge inf\{\sigma_i(x_iy_i):i \in \mathbf{N}\} \wedge inf\{\sigma_i(x_iy_iz_i):i \in \mathbf{N}\} = s \wedge t \wedge r \wedge inf\{\sigma_i(y_iy_iz_i):i \in \mathbf{N}\}$, since σ_i is order preserving. Now $(s,x)[(t,y)(r,z)] = (s,x)(w,yz) = (q,xyz)$, where $w = t \wedge r \wedge inf\{\sigma_i(y_iz_i):i \in \mathbf{N}\}$ and $q = s \wedge w \wedge inf\{\sigma_i(x_iy_iz_i):i \in \mathbf{N}\} = s \wedge t \wedge r \wedge inf\{\sigma_i(y_iz_i):i \in \mathbf{N}\} \wedge inf\{\sigma_i(x_iy_iz_i):i \in \mathbf{N}\} = s \wedge t \wedge r \wedge inf\{\sigma_i(x_iy_iz_i)\}$, and $[(s,x)(t,y)](r,z) = (s,x)[(t,y)(r,z)]$.

For $(s,x) \in T$, we have $(s,x)(s,x) = (u,xx) = (u,x)$, where $u = s \wedge s \wedge inf\{\sigma_i(x_i);i \in \mathbf{N}\} = s \wedge inf\{\sigma_i(x_i):i \in \mathbf{N}\}$. Since $(s,x) \in T$, $s \leq \sigma_i(x_i)$ for each $i \in \mathbf{N}$ and hence $u = s$. We obtain that $(s,x)^2 = (s,x)$, and T is a semilattice.

It remains only to show that multiplication on T is continuous. Let (s,x) and (t,y) be in T. We will show that multiplication is continuous at $((s,x), (t,y))$. In view of 2.15, we can assume that $(t,y) \leq (s,x)$.

We first consider the case that $0 < t,s < \infty$. For $n \in \mathbf{N}$ and $0 < \varepsilon$, let $3\varepsilon < t$ and let $W = \{(u,z) \in T: t-3\varepsilon < u < t + 3\varepsilon$ and $y_i = z_i$ for $i \leq n\}$. Then W is a basic open neighborhood of y. Let $q \in \mathbf{N}$ as guaranteed by 2.17 for $\tau = t - \varepsilon$ and ε. Now let $m = max\{n,q\}$ and define neighborhoods U and V of x and y respectively as follows:

$$U = \{(s'.a) \in T: s-\varepsilon < s' < s + \varepsilon, a_i = x_i; i \leq m\}$$

and

$$V = \{(t'.b) \in T: t-\varepsilon < t' < t + \varepsilon, b_i = y_i; i \leq m\}$$

To complete the proof, we show that $UV \subseteq W$.

Let $(s',a) \in U$ and $(t',b) \in V$. Then $(s',a)(t',b) = (u,ab)$, where $u = s' \wedge t' \wedge inf\{\sigma_i(a_ib_i):i \in \mathbf{N}\}$. We obtain that $u \leq t' \leq t + 3\varepsilon$ and $a_ib_i = x_iy_i = y_i$ for $i \leq n$, since $n \leq m$. Now $(t',b) \in V$, so that $t-\varepsilon < t' \leq \sigma_i(a_i)$ for $i \in \mathbf{N}$. If $i \leq m$, then $t-2\varepsilon < t-\varepsilon < \sigma_i(b_i) = \sigma_i(a_ib_i)$. If $m < i$, then $q < i$ and $(t-\varepsilon) - \varepsilon < \sigma_i(a_ib_i)$ from 2.17. Hence $t-3\varepsilon < t-2\varepsilon \leq s' \wedge t' \wedge inf\{\sigma_i(a_ib_i):i \in \mathbf{N}\} = u$, and $(u, ab) \in W$. ∎

Since T is homeomorphic to a closed subset of the cartesian product of the Cantor set and the unit interval, we see that T is one dimensional. Observe that $1 = (\infty, (1,1, \ldots))$ is an identity for T.

2.20 Lemma   If A is a subsemilattice of T with $1 \in A^\circ$, then $A \cap (0 \times \Pi_{i=1}^\infty S_i) \neq \square$.

Proof: At 1, T has a basis of open sets of the form: $U_j = \{(r,x) \in T : j < r, x_i = (1,1, \ldots ,1) \text{ for } i \leq j\}$ for $j \in \mathbf{N}$. Assume that j is chosen so that $U_j \subseteq A^\circ$. Define $Q = \{(j + 1,x) \in K : x_i = (1,1, \ldots ,1)$ for $i \neq j + 1$ and x_{j+1} has one zero entry$\}$. Then Q has $s(j + 1)$ elements. For each $(j + 1,x) \in Q$ we have $inf\{\sigma_i(x_i) : i \in \mathbf{N}\} = \sigma_{j+1}(x_{j+1}) = j + 1$, and hence $Q \subseteq T$, so that $Q \subseteq U$. Let $(t,z) = inf\ Q$. Since A is a subsemilattice $(t,z) \in A$. From $(t,z) \in Q$, we have $t \leq \sigma_{j+1}(z_{j+1}) = 0$ since z_{j+1} has all zero entries. Hence $t = 0$.   ∎

Let $I = \{(0,x) \in T\}$. Then I is a closed ideal of T. Let $S = T/I$, and $\pi : T \to S$ natural.

2.21 Example S is a compact non-Lawson semilattice.

Proof: That S is a compact semilattice follows from the preceding lemmas. Assume that there exists a continuous homomorphism $\phi : S \to I_m$ which is nontrivial. Then $\phi(0) < \phi(1)$, and $\phi\pi : T \to I_m$ is a continuous homomorphism such that $\phi\pi(I) = \phi(0)$. Choose r so that $\phi(0) < r < \phi(1)$. Then $(\phi\pi)^{-1}[r,\phi(1)]$ is a neighborhood of 1 in T, a subsemilattice of T, and misses I. This contradicts 2.20, and hence no nontrivial continuous homomorphism exists.   ∎

In Anderson and Hunter [1969] the following question was posed: Is the Bohr compactification of a discrete semilattice totally disconnected? J. D. Lawson showed that this need not always be the case (unpublished). We present Lawson's argument:

Let S be a compact non-Lawson semilattice and let S_d be the semilattice S given the discrete topology. Let $\phi : S_d \to S$ be the identity function on S, and let $\beta : S_d \to B(S_d)$ denote the Bohr compactification of S_d. Then there exists a continuous surmorphism $\bar{\phi}:B(S_d) \to S$ such that the diagram

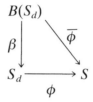

commutes. If $B(S_d)$ were totally disconnected, then it would be a Lawson semilattice from 2.9, and hence S would be a Lawson semilattice from 2.7.

ARC CHAINS

2.22 Theorem Let S be a compact connected monoid which is not a group and let V be an open set such that $1 \in V$. Then V contains an arc.

Proof: If there exists an open set W containing 1 such that $W \cap V \cap E$ has no local minimum so that V contains an arc (see Appendix B.9). Otherwise, there exists $e \in E \cap V \cap (S\backslash M(S))$ such that $eSe \cap E \cap V = \{e\}$. Since $e \notin M(S)$, eSe is nondegenerate. Hence eSe is compact connected monoid in which there is an open set containing e but not other idempotent. From 5.13 of Carruth, Hildebrant, and Koch [1983], there is a one-parameter semigroup in $eSe \cap V$ and hence it contains an arc. ∎

2.23 Theorem Let S be a compact connected semilattice. Then each element of S lies on an arc chain containing 0.

Proof: For $e \in S$, $\downarrow e = Se$ and hence $\downarrow e$ is connected. The result now follows from Appendix B.9. ∎

2.24 Lemma Let S be a locally compact Lawson semilattice and let A be a compact subset of S. Then there exists a compact subsemilattice T of S such that $A \subseteq T$.

Proof: For each $x \in A$, there exists a compact neighborhood K_x of x such that K_x is a subsemilattice. Since A is compact, some finite collection of K_x's have interiors which cover A, i.e., $K_1°, \ldots, K_n°$ cover A. Let $K = \bigcup_{i=1}^{n} K_i$. K^n is a compact subsemilattice such that $A \subseteq K \subseteq K^n$.   ∎

2.25   Theorem   Let S be a locally compact Lawson semilattice with zero 0. If S is arcwise connected, or if S is connected and locally connected, then each element of S lies on an arc chain containing 0.

Proof: Let $x \in S$. Then either hypothesis guarantees the existence of a continuum P with $0, x \in P$. Let $T = (\overline{\bigcup_{n \in \mathbf{N}} P^n})$. Then T is a compact subsemilattice of S and T is connected. From 2.23, x lies on an arc chain containing 0.   ∎

CONTINUOUS SEMILATTICES

A large portion of the effort in semilattice theory in the past decade has been exerted in the field of continuos semilattices. We present here a brief view of some of the basic concepts of this theory along with the result which is called The Fundamental Theorem of Compact Semilattices in Gierz, Hofmann, Keimel, Lawson, Mislove, and Scott [1980; see 2.29]. The work of Hofmann and Stralka in 1973 appears to have marked the merging of semigroup concepts with those of Scott to begin this era of investigation.

If S is a semilattice and $a, b \in S$ then a is said to be *way below b* (denoted $a \ll b$) provided that for each directed subset D of S, if $b \leq sup\ D$, then there exists $d \in D$ such that $a \leq d$. For $x \in S$, we denote $\Downarrow x = \{a \in S : a \ll x\}$ and $\Uparrow x = \{a \in S : x \ll a\}$.

A semilattice S is said to be *up-complete* if each up-directed set has a *sup* in S.

A semilattice S is called a *continuous semilattice* provided that S is up-complete and for each $x \in S$, $\Downarrow x$ is directed and $x = sup\ \Downarrow x$.

The *Lawson topology* on a semilattice S is the topology on S